Big Nose BIG CITY

Written by George Edward Pakenham

Illustrated by Priscila Mandryk

To my brother and all my sisters

We shared a wonderful childhood

TOGETHER

- George Edward Pakenham

Big Nose Big City is inspired by the documentary film 'Idle Threat'.
A chronicle of one mans resilient struggle to reduce air pollution in New York City
and battle global warming.

www.idlethreatmovie.com

Big Nose Big City

Hi! My name is Cyrano the bloodhound.

I have long droopy ears and a big wet nose.

All day long I sniff interesting smells.

That is how I have fun!

My owner is Giselle.

I'm so happy when she comes home from school.

Most days after class, she takes me for walks in the big park.

There are many beautiful scents.

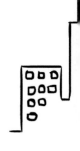

Often I see dog walkers with many, many other dogs.

Sometimes we all go to the dog park

to play, run and wrestle.

Giselle doesn't just walk me! Giselle is a dog walker.

She also takes care of my two best friends in the whole world.

One friend is Lucy the dachshund. She has tiny feet and a big bark.

My other friend is Cutie the golden lab.

I like Cutie, she is pretty!

Giselle often stops at the **Bakery** to get coffee and a pastry.

Sometimes we get treats too!!!

There are good smells near the Bakery.

When Giselle goes inside the bakery to buy our treats,

she ties us up, either to a rail...

or to a tree next to the curb.

The curbside is sometimes bad for Cutie, Lucy and me.

Often cars will park by the curb and the drivers leave the engine running.

The fumes from the exhaust pipe are awful...

right at nose level.

One day a street cleaner parked next to the curb. The tailpipe of the truck

was right near my nose and it smelled terrible!

Lucy said, "A street cleaner is supposed to help clean up the city.

He is making dirty air. It is making me sick!"

Cutie asked, "Why do drivers leave

their engines on and go nowhere?"

The tailpipe smell hurt my big wet nose.

It made it twitch. I sneezed!

Then a mother with a baby carriage approached the bakery,

stopped and started talking to a friend.

The baby carriage was near

the tailpipe of the truck.

Soon the baby started to cry.

The mother did not know why,

but I knew!

Giselle took us home that evening.

My nose was scratchy. I had sniffles.

Was it from the tailpipe fumes?

I think so!

Nearly every time we went to the bakery,

drivers parked nearby and left their engines running.

Bad smells are the worst thing for a bloodhound with a big wet nose.

One day, Giselle took me into her car.

"We are going on a play date," she said.

"But first, I want to go to buy my favorite pastry."

"What will I smell today?"

I thought.

She pulled up to the curb in front of the bakery and said:

"I'm going in for five minutes. Be my guard dog 'Cyran-nose',

my handsome boy." That made me smile.

I smelled coffee. I wagged my tail.

While I waited a traffic policeman approached the car.

I watched as he placed a **ticket** on the car windshield.

His badge said, **"Officer Dawkins."**

Giselle came out of the bakery

just as he finished.

"What are you doing?" She asked

"Writing you a ticket," said Officer Dawkins. "A ticket for idling

your engine for more than 1 minute in a school zone."

"$115 dollars for idling?" said Giselle looking at the ticket. "That is crazy!"

"It is not crazy," said Officer Dawkins. "In this city, it's the law."

Giselle looked at me and said

"Where am I going to find 115 dollars?"

"And I never even thought about engine idling before!

I wonder if that is why you felt sick the other day?"

On Monday, Giselle had 8 dogs on leashes. She was paid more money for walking more dogs. It was a fun day in the park with spring smells and flowers in bloom. We played tag all morning and wrestled too.

I made new friends.

Giselle stopped at the bakery on the way back home.

She tied us all up- Cutie, Lucy and me on the tree by the curb

and the others by the rail.

She went inside. Cutie fell asleep but just for a moment.

It was then that a big, noisy

garbage truck pulled up to the curb.

The driver left the engine on for one minute,

two minutes, three minutes

as he was sipping coffee.

The smell was horrible.

"I can't breathe,"

said Cutie, awake now and tugging away.

"I need a gas mask," barked Lucy.

"This is big trouble," I thought.

Giselle left the bakery. She immediately smelled the

dirty air and approached the driver.

"Excuse me Sir," she said.

"Did you know it's against the law to keep your engine idling in the city?

Please shut it off. My dogs need clean air!

My city needs clean air!"

"You are right, young lady," said the driver

with a smile, shutting off his engine.

Giselle smiled too.

"Thank you," she said.

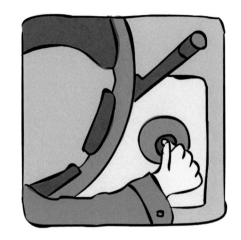

"Giselle is brave!!!" said Cutie.

Lucy said, "She saved us from the monster garbage truck."

I noticed that the bad smell suddenly went away.

"That was easy," I thought.

"All she had to do was ask!"

We continued down the street toward our home, Giselle and

all eight dogs tugging and pulling...

Some lagging behind.

But my big wet nose was held high in the air,

sniffing, sniffing and sniffing the air.

"Life was now a little better," I thought.

I knew that Giselle had changed! She had learned a lesson!

She would never let the exhaust from cars, or trucks, or buses

hurt Cutie, or Lucy, or me... ever again.

And maybe that would be good for all people who live in big cities?

But I don't know for sure.

I am just a bloodhound with a big sensitive nose.

Made in the USA
Monee, IL
27 November 2020